Forestry Commission Bulletin 76

Silvicultural Principles for Upland Restocking

P.M. Tabbush
Silviculturist,
Forestry Commission

LONDON : HER MAJESTY'S STATIONERY OFFICE

First published 1988

ISBN 0 11 710261 X
ODC 23 : (23) : (410)

Keywords: Forestry, Restocking

Enquiries relating to this publication should be addressed to the Technical Publications Officer, Forestry Commission, Forest Research Station, Alice Holt Lodge, Wrecclesham, Farnham, Surrey GU10 4LH

Contents

Silvicultural Principles for Upland Restocking

P.M. Tabbush
Silviculturist,
Forestry Commission

Introduction

Restocking is of increasing importance in British forestry, as the post-war plantings become mature and are clear felled. The Forestry Commission's annual felling programmes are expected to rise over the decade 1988/89–1998/99 from 7100 ha to 12 900 ha, while private woodland programmes, excluding broadleaves, should increase from 2600 ha to 3400 ha a year over the same period. If the second rotation is to be economically sound it is important that it is established to specified standards of stocking and uniformity at minimum cost. Excessive reductions in expenditure on planting stock, site preparation and planting can result in severe cost penalties in beating up, additional weeding, prolonged protection, loss of uniformity, lower timber quality and delayed harvesting. Managers should set standards of establishment and determine the minimum inputs necessary to achieve them in advance of these large increases in programme, and to do this they require a sound understanding of the biological processes involved.

The quality of planting stock at the time of planting is the result of nursery technique and the effects of lifting, handling, storage and transport. The performance of plants of a given quality will depend on site conditions, and clear felled sites generally offer a more testing environment than reclaimed grazing land, because stumps and slash make site preparation expensive and planting difficult and because of damage from forest insects and mammals.

Current upland restocking practice was reviewed in Forestry Commission Leaflet 84 *Guide to upland restocking practice* (Low, 1985). Recent research, both in the UK and overseas, has led to a greater understanding of the establishment process, and the intention here is to draw this body of knowledge together as a basis for the design of improved systems for upland restocking.

Plant Quality

It is self-evident that good quality planting stock is a prerequisite of successful restocking. Good quality plants are those likely to perform well under a given set of site or climatic conditions (Sutton, 1979), but it is not always easy to recognise those plant attributes which indicate or confer high performance potential.

For instance, although large plants generally perform better than small ones given adequate root:shoot ratio (Aldhous, 1950), small shoot length can confer some advantage on a site where exposure is the limiting factor (and could therefore be associated with 'high quality' on those sites). Smaller plants may also be appropriate on the more intensive forms of cultivation where weed competition is minimal. In practice, many plant attributes can be recognised as conferring high performance potential over a wide range of sites.

Morphological and physiological attributes of plant quality have been distinguished (Chavasse, 1980). Morphological factors are readily observable and are in general use for culling, grading and marketing stock. They include shoot factors such as: length, sturdiness, foliage colour (related to nutrition), degree of damage (insects, fungal infection, wilting), visual indications of 'hardness' (lignification), condition and size of the terminal bud; and root factors such as: total root length, 'branchiness' of the root system, visual assessment of root:shoot ratio, and root collar diameter.

A number of physiological factors have been shown to give a good indication of forest performance, and these include root:shoot ratio (on a dry weight or volume basis), root moisture content (Tabbush, 1987a), and root growth potential (Ritchie and Dunlap, 1980). Special equipment is usually required for their determination, and the results are not always easy to interpret, so they are usually assessed on a sampling basis.

Plate 1. Root observation boxes after 14 days in a growth chamber under test conditions in May. Left to right: placed in cold store (1°C) in mid-November, mid-December, mid-January; freshly lifted in mid-May.

Perhaps the most useful index of physiological status or vitality is root growth potential (RGP) which is an assessment of the ability to produce new roots under favourable conditions. Initial survival of planted trees depends on their readiness to produce new roots and establish intimate soil contact and so facilitate water uptake and prevent water stress (Sands, 1984).

RGP has been measured by growing the plants in moist peat in root observation boxes (Plate 1) for 14 days under favourable growing conditions, with a soil temperature of 20°C. RGP is then the number per plant of white extending roots >1 cm in length. A number of similar definitions are possible (Burdett, 1979). A chamber with a controlled environment is needed to obtain absolute values which can be compared between dates, but a number of treatments can be compared simultaneously using relative values obtained in a glasshouse or even a warm office. Experience has shown that relativities in RGP between treatments are maintained over a wide range of test temperatures (Figure 7).

It is now well established that RGP is a key factor in determining early survival and growth (Ritchie and Dunlap, 1980; Burdett *et al.*, 1983), but the relationship between RGP and plant performance is not quite straightforward. It is well known that plants can hang on to life with only one extending root, and indeed this first root is more important than the next, and so on, extra extending roots conferring very little survival advantage on plants which already have a high RGP. Conversely, plants with a low RGP will suffer reduced survival with only slight reductions in RGP. Furthermore, the survival of plants of a given RGP will be affected by site conditions, in particular dry soils (Stone and Jenkinson, 1969), cold soils and compacted soils reduce early root growth following planting.

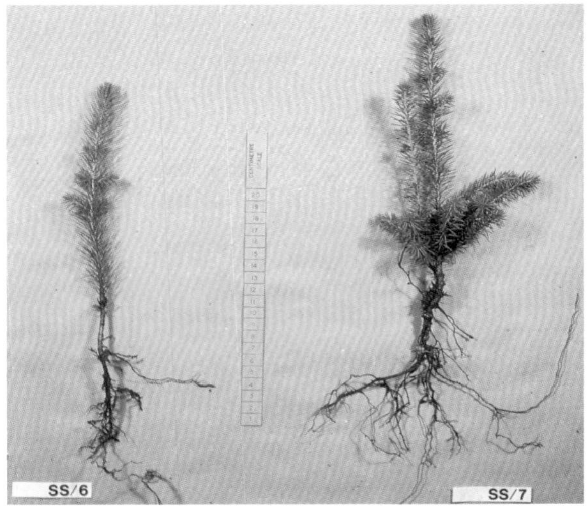

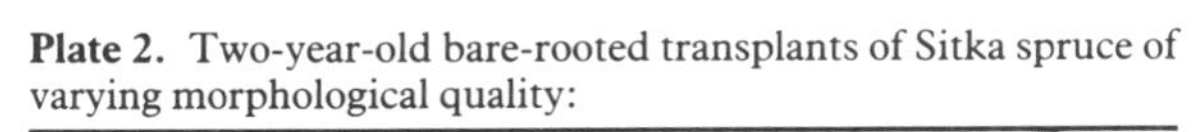

Plate 2. Two-year-old bare-rooted transplants of Sitka spruce of varying morphological quality:

Plant number	Root collar diameter (mm)	Height (cm)	Root:shoot ratio	Comment
SS4	3	26	0.2	Root:shoot ratio too low
SS6	3	21	0.5	Poorly branched
SS7	4	22	0.3	Roots not very fibrous – low root: shoot ratio
SS8	2	11	0.1	Obvious cull
SS9	4	23	0.3	Poor straggly root system
SS10	6	40	0.7	Good plant – will need careful planting

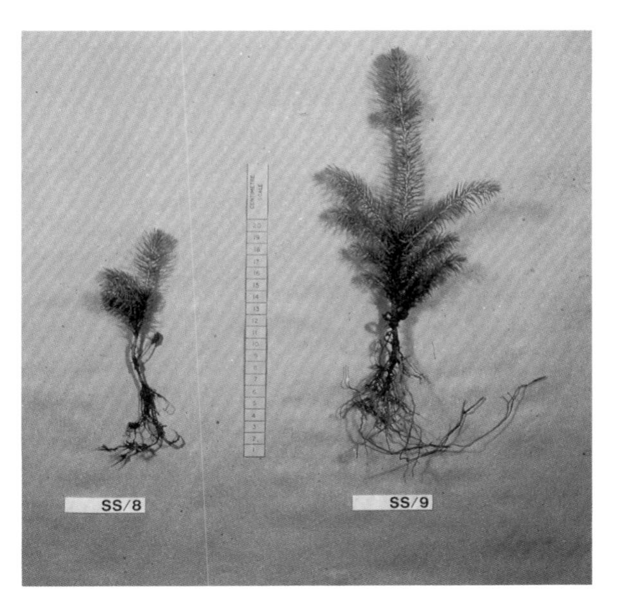

RGP, then, does not by itself give an accurate prediction of forest performance, rather plants of a high RGP are able to succeed under harsher conditions than plants with a low RGP.

Quality standards

Since clear felled sites are generally more hostile than newly ploughed pasture in terms of tree establishment, somewhat higher standards of plant quality are required than those given principally for new planting (Forestry Commission Bulletin 43 *Nursery practice*, 1st edition 1972, p.4). Table 1 gives recommended size standards for culling and grading plants intended for restocking, and some samples have been assessed and photographed for illustration (Plate 2). The ratio of root collar diameter to stem length, ('sturdiness') provides an indication of root:shoot ratio, but plants with obviously poor root systems, or which are deformed or damaged should be rejected (see illustrations). Other important morphological standards include large buds with the potential to produce strong new shoot growth in the first year, good foliage colour and absence of dead or bruised tissue.

Table 1. Stem length ranges and corresponding minimum root collar diameters for restocking plants

Species	Stem length (cm)	Minimum root collar diameter (mm)
Sitka spruce, Douglas fir, Larches	min 15	3
	max 40	7
Lodgepole pine	min 10	3
	max 35	8
Scots pine	min 10	3
	max 35	9
Broadleaves	min 20	5

Plant quality would be improved if grading could also take account of physiological criteria, but these are normally only sampled. RGP tests carried out in a glasshouse or similar environment are affected by the prevailing weather, and can only be used for comparison amongst batches of plants tested simultaneously. Plants will often be available for comparison, e.g. which have not been subjected to the treatment under test. Where no useful comparison can be drawn, it is only possible to specify that RGP should be greater than zero at planting. Where growth chambers providing a controlled environment are available, it can be specified that with a root temperature of 20°C, 16 hour days and high light intensity, the RGP of Sitka spruce after 14 days should be >10 new roots >1 cm long per plant.

Fine root moisture content expressed as a percentage of dry weight (RMC) should always exceed 250 per cent in Sitka spruce and 300 per cent in Douglas fir (Tabbush, 1987a). Once RMC has fallen below these levels the damage inflicted cannot be rectified by rewetting; irrigation or soaking is only of service in maintaining RMC above critically damaging levels. Measurement of RMC requires a drying oven and accurate (0.01 g) balance, but such facilities could be provided at centralised locations with technical staff, e.g. large nurseries. This would also enable measurement of root:shoot ratio on a dry weight basis, which for Sitka spruce should exceed 0.3.

Plant Handling

Plant handling damage may result from desiccation, mechanical shock, extremes of temperature or exposure to chemicals (usually insecticides). Successful establishment is achieved by preventing this loss of plant potential from reaching critical levels.

Desiccation

Bare-rooted plants are usually protected from drying conditions by packaging in air-tight containers – polythene bags or waterproof boxes. Fine roots are easily damaged by desiccation; and Hermann (1967) for example, recorded damage to Douglas fir seedlings with washed roots after only one minute's exposure to desiccating conditions in a growth chamber. In experiments at Wykeham, Yorkshire (Tabbush, 1987a) Douglas fir was more susceptible than Sitka spruce, but survival was reduced significantly in the latter species after exposure to windy conditions in March for 70 minutes. Roots which retain a covering of nursery soil dry out more slowly, but excess nursery soil can result in soil getting on to foliage and serving as a source of fungal infection during storage. Dipping roots in water-retentive substances such as sodium alginate protects against drying (Brown, 1972), but is in itself an additional plant handling operation which can result in mechanical injury.

Root desiccation reduces RGP. Furthermore, desiccated plants are especially vulnerable to rough handling (Tabbush, 1986a). Most forms of handling damage are more severe during the period of active growth (Ritchie, 1986), and in general, resistance to stress is high when RGP is high (December–March).

The risk of desiccation is great during storage, and bags or boxes must be completely sealed to prevent the escape of moist air. Even small slits in bags have resulted in deaths in directly refrigerated cold stores.

Physical shock

Rough handling has been simulated experimentally by dropping sealed bags of plants, roots downwards, 1–15 times from a height of 3 m on to a hard floor. Although no obvious physical damage was visible, the treatment resulted in reductions in RGP, survival, growth and mycorrhizal development (Tabbush, 1986a; 1986b). There is evidence that physiological changes take place as a result of mechanical damage to roots (Coutts, 1980) and it may be that the large reduction in root development observed is the result of some physiological change.

Soft packages such as polythene bags are easily compressed and the plants within them bruised. Racking or shelving can be employed to reduce this but it is expensive and requires extra space. Rigid containers (e.g. plastic or waterproof cardboard boxes) may offer some protection, and may be more suitable for mechanised handling.

A high level of supervision is required to ensure that plants are handled carefully, and this is not always feasible at remote locations. Future systems should involve mechanical handling where possible, and the use of rigid containers and stacking systems could well reduce handling and transport costs whilst minimising the opportunity for crushing and mechanical shock.

Extremes of temperature

Tissue damage by heating is the result of both the temperature reached and the length of time for which it is sustained. In general temperatures in excess of 35–40°C are likely to cause direct damage (Levitt, 1972). Such temperatures have been reached after about 1 hour when clear polythene bags of plants have been left in direct sunlight (Tabbush, 1987b). Heating may also occur by bacterial and plant respiration if plants are too tightly stacked so that heat cannot be removed by freely circulating air. In storage each package should have at least one side exposed, and, for long-term storage, stacks of plants should not exceed 1 m in depth.

High temperatures which are insufficient to cause direct damage may result in the hastening of growth processes, and this can result in a lowering of resistance to stress so that plants become more difficult to handle without damage. It is therefore important to keep plants cool at all times.

Low temperature damage can be inflicted by cold storage on plants not physiologically conditioned to withstand it. Severe winter cold can have a similar effect on plants stored out of doors. Conifer shoots generally become tolerant of temperatures below −40°C in January–February, but may be killed by temperatures just below freezing in June. Roots, on the other hand, are never able to tolerate temperatures much below −25°C (Studer *et al.*, 1978) and so can be damaged by severe winter cold when stored without the protection of the soil, for example stored in the open in bags or as containerised stock. Temperature extremes are reduced in sheltered situations, e.g. under a dense canopy of trees.

In the British climate, damage is often inflicted by unseasonal weather, especially a warm spell in spring followed by a reversion to frosty weather. Plants may be stored most safely in a well managed cold store.

Storage

It is often necessary to store plants for short periods between handling and treatment operations (e.g. culling, grading or treatment with insecticide), or for longer periods between lifting and planting. The flow of plants through the system must be organised so as to reduce the opportunity for the types of handling damage described above.

Cold storage

Cold stores offer nursery and forest managers great flexibility. Once plants have become fully 'storable' in late autumn thay may be lifted to suit nursery soil conditions, and the availability of labour, and put into store unconstrained by weather conditions or programme considerations at the forest. They may, under some circumstances, be culled, graded and returned to the store for a period before despatch to the planting site at a time when forest soil conditions are suitable. Plants can be held in store where they will keep a high RGP and inactive buds well beyond the time in spring when freshly lifted stock has active buds and RGP has decreased (Plate 1). Thus trees from the cold store with a high survival potential can be planted in late spring or early summer when forest soil conditions are favourable.

However, serious damage can be inflicted if plants are put into store before they become inactive in autumn, or after they become active in spring (e.g. Duryea and McClain, 1983). It is not clear whether it is bud dormancy, frost hardiness, RGP or some other factor which determines storability. These factors are interrelated, but they are not causally related, and can vary independently. In experiments at the Northern Research Station (Figures 1 and 2) the RGP of Sitka spruce rose or was maintained in cold storage whilst that of Douglas fir declined. For upland nurseries Sitka spruce is the most appropriate species for cold storage.

Most cold stores are directly refrigerated, i.e. the cooling unit and fan are within the storage chamber. This has a desiccating effect on the atmosphere inside the store, and plants must therefore be held inside sealed waterproof containers. In contrast, in a humidified store, air is blown across cold water and the resultant cold, moist air is ducted to all parts of the storage chamber. The temperature cannot be reduced to 0°C or below. In directly refrigerated stores the temperature should be selected within the range +2 to −2°C and maintained within 1°C of the selected figure. It should not be allowed to fluctuate above and below 0°C. In humidified stores the temperature should not exceed +2°C, and the relative humidity should be maintained above 95 per cent.

If plants are stored at +2°C in either type of store there is less risk of low temperature damage to plants which are not fully storable, either because they have been put into the store before they have become fully storable or because they have 'dehardened' within the store in late season. On the other hand, there is more risk of infection by harmful fungi than at −2°C. Furthermore, respiration is slowed at lower temperatures, and food reserves are burned up more slowly. Provided plants are in a fully storable condition, long-term storage is probably safest at −2°C in a directly refrigerated store.

The risk of mould developing on plants during storage is increased by over-filling shelves or bags and by over-tightening bundles, and by the presence of plants with damaged, wet or dirty shoots or foliage, or weeds and other plant debris. Humidified stores should be washed out thoroughly using hot water containing mould inhibitor or disinfectant before re-use. In directly cooled stores the plants are inside sealed containers, and theoretically there should be no need for more than general cleanliness. However, if mould grows on timber shelves etc., the cleaning procedure for humidified stores should be followed.

Plate 3. 'Canopy storage' of Sitka spruce transplants in coextruded black and white polythene bags. The bags are sealed to prevent moisture loss, and are reflective and opaque to prevent overheating.

Temporary storage

Temporary storage is often necessary, e.g. between delivery and planting at the forest, or pending treatment at the nursery or depot. In the past 'heeling in' or 'sheughing' has been commonplace, but since this involved a good deal of plant disturbance, and since soil conditions which afford proper protection to the roots are rare, it is best avoided. Plants can be stored in coextruded black/white polythene bags (Tabbush, 1987b) in a cool, shaded environment under a dense tree canopy, on the north side of a building, or in a well ventilated shed ('canopy storage') (Plate 3). Sitka spruce will normally tolerate these conditions from December to mid-April. Douglas fir deteriorates easily, however, and should not be stored in this way for more than 6 weeks during the period mid-January to mid-April (Figure 4).

Timing

Seasonal variation in RGP

The response to lifting, handling and planting is determined to a large extent by the physiological condition of the plant, and this varies as the season progresses.

Dormancy is induced in autumn in response to declining day length and temperature. In some species, e.g. Douglas fir, there is evidence that water deficits in late summer play an important role in the induction of dormancy (Ritchie and Dunlap, 1980). Full dormancy is attained in October–December, depending on the season, and from then on, dormancy is released progressively as the 'chilling requirement' is met by accumulated winter cold. During this period RGP rises steeply. The variation in RGP for Sitka spruce and Douglas fir when lifted from the open nursery on successive dates was assessed in an experiment at the Northern Research Station in 1985/86. The steep rise in RGP reached a peak in Sitka spruce (Figure 1), just before the buds began to swell in spring. In contrast, the peak for Douglas fir was attained in December–January (Figure 2). By April–May, buds began to burst in both species and RGP was low. Planting after the buds begin to develop makes the plants vulnerable, since it is difficult for the slowly developing root system to meet the demands of the rapidly developing foliage (Plate 1).

The normal pattern of seasonal variation in Sitka spruce is shown in Figure 3. Frost hardiness is measured by determining the temperature required to kill shoot tissue by freezing in a programmable chamber. Dormancy is measured by subjecting plants to

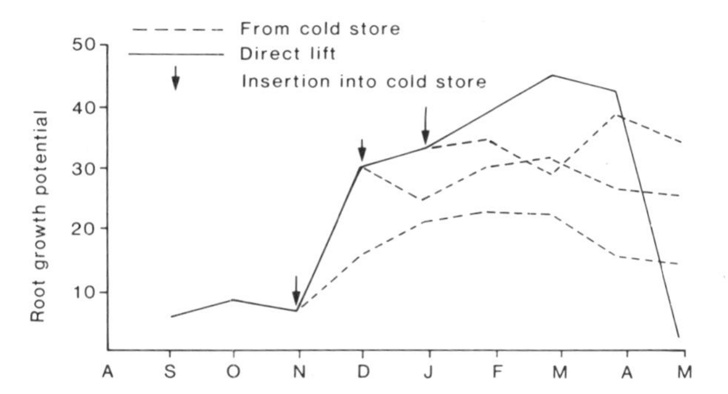

Figure 1. Seasonal changes in the root growth potential of Sitka spruce either directly lifted from the open nursery or after insertion into cold store in mid November, December and January.

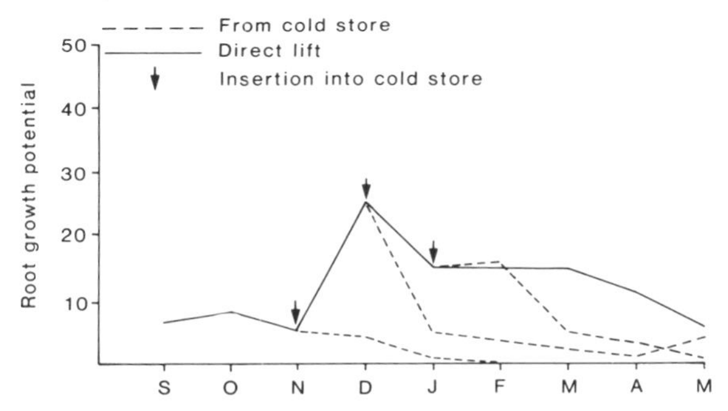

Figure 2. Seasonal changes in the root growth potential of Douglas fir either directly lifted from the open nursery or after insertion into cold store in mid November, December and January.

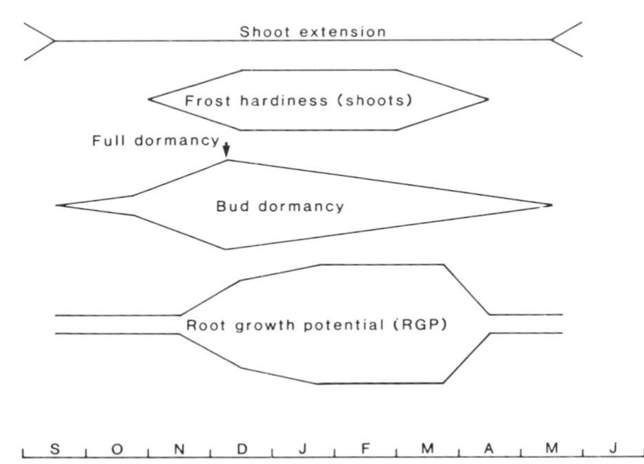

Figure 3. Diagrammatic representation of seasonal changes in RGP, bud dormancy, frost hardiness and shoot extension in Sitka spruce.

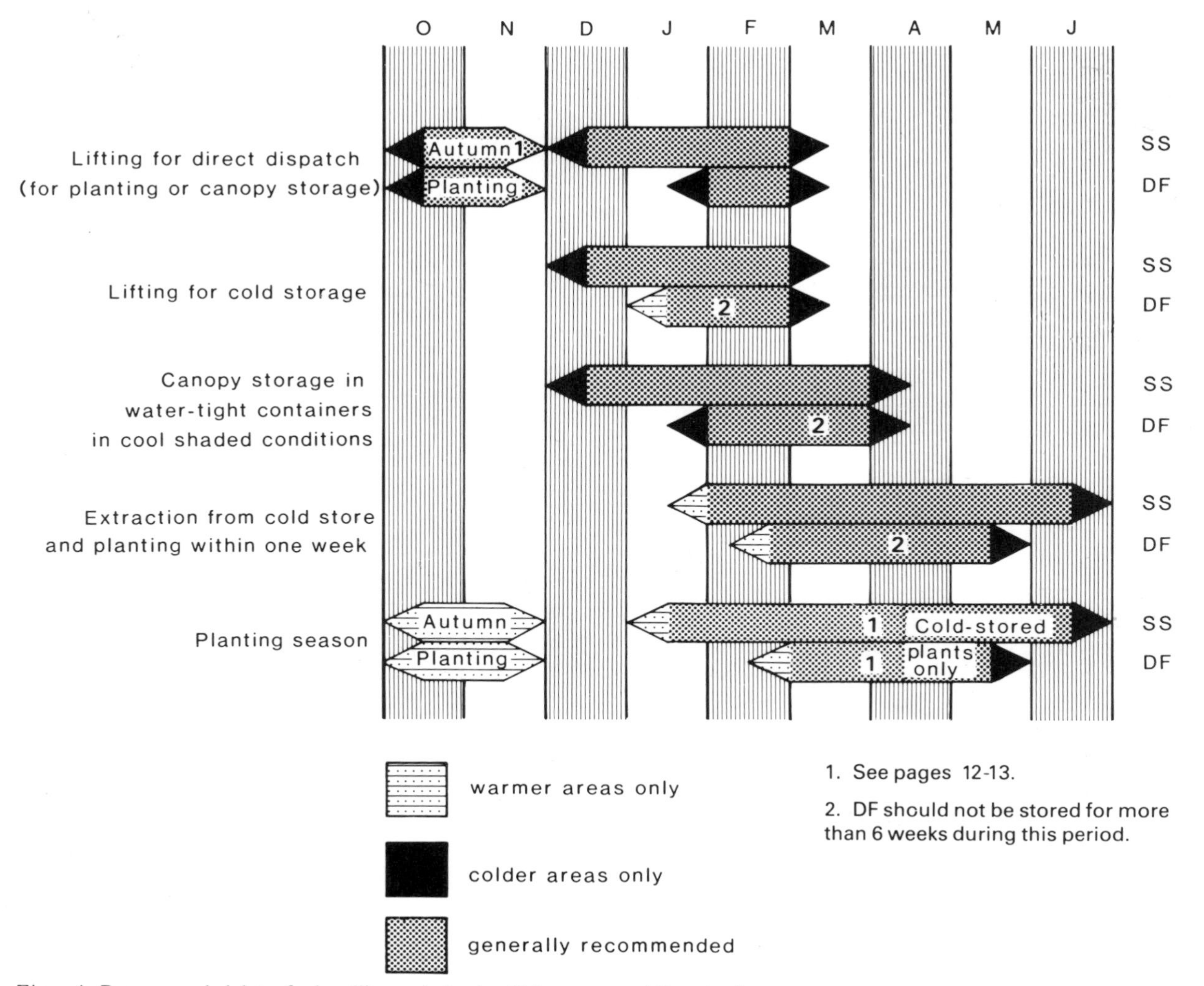

Figure 4. Recommended dates for handling and planting Sitka spruce and Douglas fir.

long days and warm temperatures and counting the number of days before budburst. Neither frost hardiness nor bud dormancy gives a clear indication of RGP. Direct measurement of RGP appears to give the best guide to lifting date, since the change in autumn is relatively abrupt and easy to detect, and RGP has been related to resistance to stress (Ritchie, 1986). Thus lifting for cold storage should not start until RGP has begun to rise. From the point in spring at which RGP falls, only cold stored stock should be used. The dates will vary depending on species and season, and the cycle may be manipulated to some extent by nursery techniques. For example, repeated root wrenching can advance the date at which RGP starts to rise in autumn. Undercutting regimes, such as those prescribed for precision sown and undercut stock (Mason, 1986) can improve the storability and RGP of both Douglas fir and Sitka spruce, and manipulation of day length, fertiliser and irrigation regimes can have the same effect on container-grown stock (Burdett, personal communication).

Planting dates

Planting dates should be chosen to ensure that roots will grow rapidly after planting and before budburst. Low levels of RGP correspond to low levels of resistance to desiccation or rough-handling (Ritchie, 1986), that is, plants suffer more damage from a given handling treatment when RGP is low. There is therefore an advantage in organising the plant supply system so that lifting, handling and planting occur

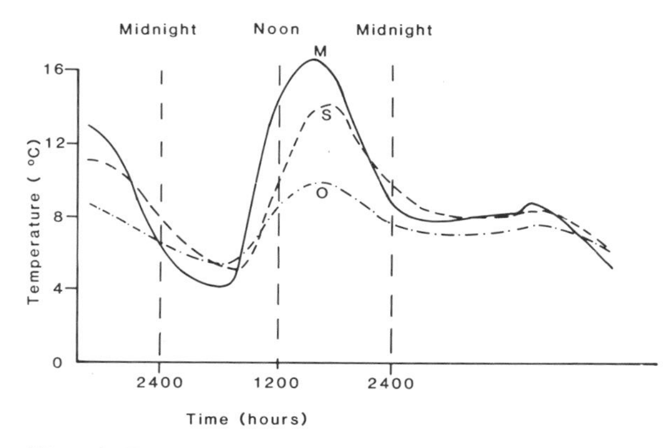

Figure 5. Temperature 5 cm below the surface (means of four replicates) in M – mounds; S – scrapes; O – undisturbed ground (from an experiment on a peaty gley soil at Kershope Forest, Cumbria). Mean of eight mounds measured in April.

when RGP is high. On the other hand, root growth will only occur in the forest if soil temperature is reasonably high, for example with significant parts of the day when the temperature 5 cm below the soil surface exceeds 5°C, and so lifting and planting dates must be chosen with both RGP and soil conditions in mind.

Recommendations for the timing of plant handling, storage and planting operations for Sitka spruce and Douglas fir are summarised in Figure 4.

Seasons for plant handling are more restricted in the south and at lower elevations where the growing season is longer. For example, 'canopy storage' should normally finish by the end of March in Wales and southern England. Planting, on the other hand, can take place earlier in spring at warmer locations. Autumn planting is most appropriate where soils are likely to remain warm into November. Extensions for warmer or cooler areas are indicated in Figure 4.

Plants may be lifted for autumn planting from early October until the end of November provided shoots have become fully lignified ('hardened'). From then on, cold soil will limit establishment success. Because it tends to be planted on cooler sites, Sitka spruce is generally less suitable for autumn planting than Douglas fir. Plants destined for autumn planting should be identified by early August so that nurserymen can condition them, for example by undercutting to prevent soft extension growth in late season.

In general, planting after the end of March should use cold stored plants only. However, if plants are canopy stored at a northern location at high elevation, bud break will be delayed, and the period before switching to cold stored plants may be extended until mid-April.

Unseasonal weather conditions will affect safe lifting dates for cold storage, thus a mild February might result in plants being too active to be placed in store in March, and conversely a cold October might enable cold storage of Sitka spruce from early November. Cold storage of Douglas fir should be for a maximum of 6 weeks during the period between January and the end of May.

After extraction from cold store, plants will remain inactive for a short period during which resistance to stress is high. Programmes should be organised to allow cold store plants to be planted within a week of emergence from store, and this is particularly important after the end of March and at warmer localities.

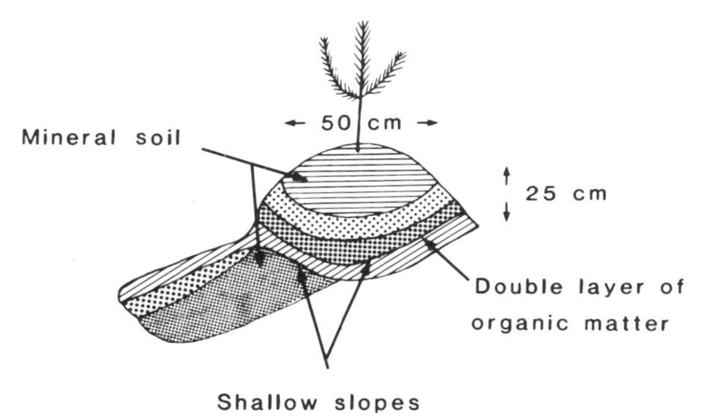

Figure 6. Diagrammatic representation of an ideal mound for planting.

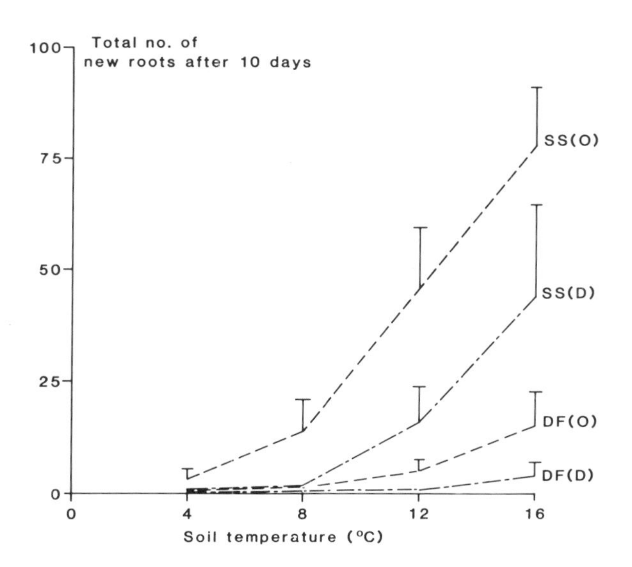

Figure 7. Number of new roots produced at different soil temperatures by Sitka spruce (SS) and Douglas fir (DF) after: O – careful handling; D – dropping in sealed polythene bags 10 times from a height of 3 m.

Plate 4. Double-mouldboard restocking plough (D60/T90/t) trailed by a D6 Caterpillar tractor fitted with apical-track shoes for flotation on boggy ground.

Site Preparation

Although the long-term growth and stability of a stand may be affected by the ground configuration and drainage of the site as a whole, early survival and growth are affected only by the microsite conditions immediately surrounding the newly-planted tree. The root system will develop most readily in warm, moist soil, with an adequate supply of mineral nutrients, and with no competing vegetation. Removal of the layer of litter and humus covering the mineral soil improves the soil temperature regime, and this effect increases if a mound or ridge is formed (Figure 5). Root growth is minimal below 4°C (Figure 7) and increases in response to the number of degree-hours above this critical soil temperature. More heat is absorbed by the bare mineral soil than by litter or a grassy surface, and this heat is available at night to increase the air temperature around the newly planted

Plate 5. Leno patch-scarifier mounted on an agricultural tractor.

Plate 6. Bräcke patch-scarifier trailed by Timberjack 240 skidder.

tree, and so reduce the risk of frost damage. Ploughing and scarifying have been shown to reduce the incidence of damage by the large pine weevil (*Hylobius abietis* L.).

In most upland situations the planting position should be raised above the original ground surface to give local drainage and maximum improvement in the temperature regime. These conditions also promote the mineralisation of nitrogen. The only exception is that on very dry soils (e.g. littoral sands) planting should be in the 'scrape' position, below the original ground surface.

Experiments have shown continually improving microsite temperature up to a mound height of 50 cm, but it will prove difficult for trees to develop an evenly spread, and therefore supportive, root system in tall, narrow mounds. In practice, a mound approximately 25 cm tall and with shallowly sloping sides so that it is at least 50 cm in diameter is recommended. The layer of decomposing organic matter should be readily available to the plant, especially on cold upland soils with low rates of nitrogen mineralisation (Figure 6).

Mounds are particularly appropriate in areas of high windthrow hazard, since the production of close-spaced furrows, which restrict root spread, is avoided. However, raised microsites can be produced by disc trenchers or even ploughs on the better, drier soils.

As soil temperature rises over the range 4–16°C, there is an increasingly strong root response (Figure 7). Plants with a low RGP, either because of rough treatment or because of species (e.g. Douglas fir), are especially likely to benefit from the improvement in soil temperature that results from cultivation. Conversely, stock with a high RGP will survive and grow, given moderately favourable site conditions, without expensive site preparation.

Machinery

Rectification of wrongly laid out first rotation ploughing, or fracturing of an ironpan or indurated layer can be accomplished with a double-mouldboard restock-

Plate 7. TTS–Delta disc-trencher mounted on a modified Timberjack 380 skidder with supplementary engine.

ing plough (D60/T90) (Plate 4) which ploughs deep enough to lift stumps rather than meet them head-on. The plough is usually trailed by a D6 Caterpillar tractor or mounted on a D7. The 'humpy' carriage is used as this allows easier movement of slash accumulations. The high cost of ploughing is directly related to the low output of these machines (0.1–0.2 ha h^{-1}). Ironpans or induration may be broken using a ripping tine, although again, outputs are low. Although cultivation experiments, especially on heathlands, indicated increased early growth with increasing volume of soil disturbed (Taylor, 1970), this effect seems to become reversed during the thicket stage (e.g. Wilson and Pyatt, 1984), possibly because high nitrogen mineralisation rates induced by cultivation slow down once the canopy is closed. There is no evidence of a long-term improvement in growth rate, unless a perched water table is released by the fracturing of an ironpan or indurated layer. Furthermore, there is evidence that the furrows present a serious barrier to roots, leading to early windthrow (Coutts, 1983).

Scarifiers (from the surgical term – 'to make a shallow incision') are less likely to exhibit these disadvantages and generally operate at high outputs (0.5–1.2 ha h^{-1}). The cheapest form of cultivation is to use a patch-scarifier to scrape away the slash and duff and expose mineral soil at each planting position. These machines have mattock-wheels which are either braked hydraulically ('Leno') (Plate 5) or coupled by reduction gears to a rubber-tyred ground-wheel, so that the mattock-wheel turns at about half the speed of the ground-wheel ('Bräcke') (Plate 6). They are particularly suited to sites with freely draining soil and light slash cover.

The disc-trenchers (e.g. TTS–Delta) (Plate 7) produce twin continuous cleared traces and the versions with powered discs are particularly effective in dealing with slash. They leave no clearly defined raised planting position although there is often an opportunity to plant at the edge of the scrape on a slightly raised position between the scrape and the loose ribbon of slash and spoil. Disc-trenchers are therefore most

Plate 8. Sinkkilä 2 patch-scarifier/mounder mounted on a D6 Caterpillar tractor with apical track shoes.

appropriate for freely drained soils – brown earths, podzols and freely drained littoral sands – but may be extended to the better end of the surface water gleys, or used simply for breaking and redistributing deep spruce slash on those wet soils which will support them. The angle of attack of the discs can be varied, e.g. to give a broader scrape for weed control (and frost protection) on freely drained soils liable to rapid weed invasion. The powered version can work with a broader angle than the unpowered, as the latter ceases to rotate at very broad angles.

Surface water gleys and peaty gleys on slopes greater than 3° would normally receive drainage at a minimum of 40 m spacing (Pyatt and Low, 1986), and under these circumstances there is scope for mounding by continuous-acting mounder. Currently, the only such equipment available is the Sinkkilä 2 (Plate 8), which is a patch-scarifier modified to produce mounds of about 75–100 litres which stand 20–30 cm above the original soil surface. By adjustment, the mound may be varied in size and distance from the excavation. (This equipment can also be adjusted to operate as a patch-scarifier on drier ground, but the advantages of a mound operate on all but the very driest of sites such as littoral sands.) The Bräcke mounder (Plate 9) also shows potential for mounding on a wide range of sites.

On the wettest sites, with slopes less than 3° where the recommended drain spacing is 20 m, or where stumps and ground conditions preclude the use of continuous-acting mounders, mounds can be spread at the desired spacing by an excavator at little extra cost while digging the drains.

Mounds will generally provide the opportunity for relatively unrestricted root development, and this should result in a more symmetric root system and hence greater stability. Against this, mounding ground ploughed for the first rotation can block furrows and reduce drainage.

A summary of recommended equipment according to soil type is given in Table 2. On many restocking sites, first rotation mounds or plough ridges may offer

Table 2. Recommended equipment for site preparation on clear felled sites

Soil type	Soil classification (Pyatt, 1982)	Recommended equipment*
Brown earths	1, 1d, 1u, 1z	P, D, CM
Podzols/intergrade ironpan soils	3, 3p, 4b	P, D, CM
Freely and imperfectly drained littoral soils	15e, 15i	P, D
Ironpan soils/indurated soils	4, 4p, 4z, 4zx	P, D, CM, R
Indurated gley soils	6zx, 7zx	CM, EM, R
Brown gleys and imperfectly drained brown earths	7i, 1g	D, CM, EM
Surface water and groundwater gleys	5, 6, 6p, 7, 15g,	CM, EM (D for slash treatment only)
Peats	8, 9, 10, 11	EM

**Key to equipment codes*

P = patch-scarifier, e.g. Leno, Bräcke. The Bräcke is the preferred machine because of the gently sloping profile of the scrape

D = disc-trencher, e.g. TTS Delta

CM = continuous-acting mounder, e.g. Sinkkilä 2 or Bräcke mounder

EM = discontinous-acting (excavator) mounder

R = ripping tine for breaking ironpans or induration if upper soil horizons are noticeably wetter than lower soil horizons. Generally used in conjunction with a scarifier

suitable microsites for planting and there may be no need to cultivate at all. The table is to aid equipment choice assuming site conditions have already dictated some form of site preparation.

Planting

Planting method

Study of the planting operation has been much neglected, and is complicated by the variety of site types, cultivation methods and planting stock types. A larger, more fibrous root system can be planted in a large mound of mineral soil than can be planted in a small raised position near a stump on an uncultivated site with a high water table. Compact root systems, produced by a container system or by side- and under-cutting in the open nursery, present fewer problems in planting. The structure of the root system of the mature tree, and hence tree stability, is strongly influenced by the early development of the root system after planting (Coutts, 1983), and so microsite type and planting method can have a profound impact on the economic value of the plantation. It is important to arrange roots at planting so that they can develop on all sides, and so that they are oriented downwards or horizontally. A root system stuffed into a narrow slit so that many of the roots point upwards is unlikely to result in rapid establishment and long-term stability. This is particularly important with pines, larches and Douglas fir which rarely produce adventitious roots and so are dependent for their root geometry on the distribution of the roots present at planting.

In a planting study at Glentress (Borders Region) Sitka spruce transplants planted roughly and firmed with the heel of the boot produced fewer new roots during the first growing season than trees planted gently and firmed with the ball of the foot. Gentle, careful planting will result in more reliable establishment than hurried, rough planting.

Weeding

Weeds compete with newly planted trees for light, moisture, rooting space, and nutrients, and some weeds even interfere biochemically (allelopathically) with tree growth, e.g. heather (Read, 1984) and some grasses (Maclaren, 1983).

Attempts to establish the benefits of weeding in the uplands have indicated that for Sitka spruce on many upland sites, typically with a high rainfall and impeded drainage, competition for moisture is negligible and it is only necessary to remove competition for light (Tabbush, 1984). This is an important distinction since hand cutting of weeds or mowing only removes competition for light, and may exacerbate competition for moisture, whereas chemical weed control eliminates both types of competition by killing both the aerial parts and the roots. By contrast, experiments in low rainfall areas on well drained soils show very significant effects of moisture competition, i.e. large benefits of chemical weed control and small or negative benefits from cutting or mowing (Davies, 1985, 1987).

Plate 9. Bräcke mounder trailed behind an Ösa 260 'short processor'.

However, it would be misleading to draw a sharp distinction between uplands and lowlands in this respect. Experiments are usually carried out using first class planting stock, and with small-scale, and well controlled, handling and planting. The critical time between planting and the establishment of soil contact through root development is also the critical time for weed competition. The response of the root system is modified by soil conditions, and the response to low soil water potential (dry soil) is to produce fewer roots (Stone and Jenkinson, 1969). Thus the effect of root zone competition is more likely to be felt by a plant of low RGP than by a more vigorous plant.

There will therefore be many instances where root competition does prove important even in the uplands, especially on well drained soils on southern aspects or in dry seasons. Since, in any case, chemical weeding is much cheaper than handweeding, the former method is much to be preferred. A weed-free area of at least 1 m diameter should be present at planting (i.e. by cultivation or preplanting application of herbicide), and maintained for at least the first growing season. Weeding in subsequent seasons should only be necessary on the most fertile sites. When vegetation is dense and tall, spot weeding can be inadequate and band or broadcast weed control may be necessary to prevent smothering. Detailed recommendations are given in Forestry Commission Booklet 51 *The use of herbicides in the forest* (Sale *et al.*, 1986; this Booklet is now out-of-print but the information is being updated and will form the basis of a future FC publication.)

Conclusions

In considering plantation establishment, whether cultivation, weed control, or some other aspect, one is constantly reminded of the importance of plant quality at the moment of planting. Given minimal site preparation and even a quite hostile microsite, a tree with first class morphological and physiological quality, handled well and planted with care, will at least survive and grow. Given good site preparation and weed control the same plant will respond with rapid early growth, and consequent savings on second and subsequent weedings. Conversely, substandard planting stock gives low survival percentages unless site preparation and weed control are of the first quality, and even given expensive inputs at the beginning, early growth is slow, and the manager is faced with repeated beating up and weeding which can be punitively expensive, especially on fertile sites with vigorous weed growth. Slow establishment, whether due to deaths or planting check, also results in uneven plantations and delayed harvesting.

It seems well worthwhile to pay for the highest quality stock available, and if necessary to ensure that it is truly vigorous by physiological testing, either at nursery or receiving forest. Root growth potential, root:shoot ratio and root moisture content provide the planting forester with vital information about the stock he is expected to plant. Knowledge of their variation will also provide guidance to the nursery manager – he will be able to condition stock to bring it to the ideal condition in time for planting.

It is important to raise managerial expectations; beating up should be exceptional; it is usually the result of inadequate attention to plant quality, plant handling and the timing and quality of the planting operation; sometimes it is due to inadequate microsite quality, and occasionally to inadequate measures against forest pests. The target must be clearly defined, for example 2500 stems per net hectare with no more than 2.4 m between adjacent plants at year 5. New systems of plant production and plant handling which recognise the importance of plant vitality at planting must be designed, and these new systems must be brought together with appropriate investment in site preparation and weeding. Great financial penalties are often incurred by excessive short-term economies, and insufficient attention to the cost of the entire establishment system.

Acknowledgements

The author is grateful to all the foresters and researchers who have contributed to the work summarised here. My thanks to Mr J. Williams who drew the figures, and Mrs S. M. Swan who typed the draft. Figure 3 derives from collaborative research with the Institute of Terrestrial Ecology. Dr J. J. Philipson, Mr D. B. Paterson and Mr S. A. Neustein made helpful comments at the drafting stage.

References

ALDHOUS, J.R. (1950). The effect of height and diameter on growth and survival. *Forestry Commission Report on Forest Research 1950*, 22–23. HMSO, London.

BROWN, R.M. (1972). Root dips in sodium alginate for seedlings and transplants. *Forestry Commission Report on Forest Research 1972*, 27-28. HMSO, London.

BURDETT, A.N. (1979). New methods for measuring root growth capacity: their value in assessing lodgepole pine stock quality. *Canadian Journal of Forest Research* **9**, 63–67.

BURDETT, A.N., SIMPSON, D.G. and THOMPSON, C.F. (1983). Root development and plantation establishment success. *Plant and Soil* **71**, 103–110.

CHAVASSE, C.G.R. (1980). Planting stock quality: a review of factors affecting performance. *New Zealand Journal of Forestry* **25**, 144–171.

COUTTS, M.P. (1980). Control of water loss by actively growing Sitka spruce seedlings after transplanting. *Journal of Experimental Botany* **31**, 1587–1597.

COUTTS, M.P. (1983). Development of the structural root system in Sitka spruce. *Forestry* **56**, 1–16.

DAVIES, R.J. (1985). The importance of weed control and the use of tree shelters for establishing broadleaved trees on grass dominated sites in England. *Forestry* **58**, 167–180.

DAVIES, R.J. (1987). *Trees and weeds.* Forestry Commission Handbook 2. HMSO, London.

DURYEA, M.L. and McCLAIN, K.M. (1983). Altering seedling physiology to improve reforestation success. In, *Seedling physiology and reforestation success,* eds. M.L. Duryea and G.N. Brown, 77–114. Nijhoff/Junk, Dordrecht.

HERMANN, R.K. (1967). Seasonal variation in the sensitivity of Douglas-fir seedlings to exposure of roots. *Forest Science* **13**, 140–149.

LEVITT, J. (1972). *Responses of plants to environmental stresses,* chapter 11, 229–321. Academic Press, London.

LOW, A.J. (ed.) (1985). *Guide to upland restocking practice.* Forestry Commission Leaflet 84. HMSO, London.

MACLAREN, P. (1983). Chemical warfare in the forest. *New Zealand Journal of Forestry* **28**, 73–92.

MASON, W.L. (1986). *Precision sowing and undercutting of conifers.* Research Information Note 105/86/SILN. Forestry Commission, Edinburgh.

PYATT, D.G. (1982). *Soil classification.* Research Information Note 68/82/SSN. Forestry Commission, Edinburgh.

PYATT, D.G. and LOW, A.J. (1986). *Forest drainage.* Research Information Note 103/86/ SILN. Forestry Commission, Edinburgh.

READ, D.J. (1984). Interaction between ericaceous plants and their competitors with special reference to soil toxicity. In, *Aspects of Applied Biology* **5**, *Weed control and vegetation management in forests and amenity areas,* 195–209.

RITCHIE, G.A. and DUNLAP, J.R. (1980). Root growth potential: its development and expression in forest tree seedlings. *New Zealand Journal of Forest Science* **10**, 218–248.

RITCHIE, G.A. (1986). Relationships among bud dormancy status, cold hardiness, and stress resistance in 2+0 Douglas-fir. *New Forests* **1**, 29–42.

SALE, J.S.P., TABBUSH, P.M. and LANE, P.B. (1986). *The use of herbicides in the forest.* Forestry Commission Booklet 51. Forestry Commission, Edinburgh.

SANDS, R. (1984). Transplanting stress in radiata pine. *Australian Forest Research* **14**, 67–72.

STONE, E.C. and JENKINSON, J.L. (1969). Influence of soil water on root growth capacity of Ponderosa pine transplants. *Forest Science* **16**, 230–239.

STUDER, E.J., STEPONKUS, P.L., GOOD, G.L. and WIEST, S.C. (1978). Root hardiness of container grown ornamentals. *Horticultural Science* **13**, 172–174.

SUTTON, R.F. (1979). Planting stock quality and grading. *Forest Ecology and Management* **2**, 123–132.

TABBUSH, P.M. (1984). Effects of different levels of grass-weeding on the establishment of Sitka spruce. In, *Crop protection in northern Britain 1984*, 339–344. Forestry Commission, Midlothian.

TABBUSH, P.M. (1986a). Plant handling. *Forestry Commission Report on Forest Research 1986*, 17. HMSO, London.

TABBUSH, P.M. (1986b). Rough handling, soil temperature and root development in outplanted Sitka spruce and Douglas fir. *Canadian Journal of Forest Research* **16**, 1385–1388.

TABBUSH, P.M. (1987a). Effect of desiccation on water status and forest performance of bare-rooted Sitka spruce and Douglas-fir transplants. *Forestry* **60**, 31–43.

TABBUSH, P.M. (1987b). *The use of co-extruded polythene bags for handling bare-rooted planting stock.* Research Information Note 110/87/SILN. Forestry Commission, Edinburgh.

TAYLOR, G.G.M. (1970). *Ploughing practice in the Forestry Commission.* Forestry Commission Forest Record 73. HMSO, London.

WILSON, K. and PYATT, D.G. (1984). An experiment in intensive cultivation of an upland heath. *Forestry* **57**, 117–141.

Printed in the United Kingdom for Her Majesty's Stationery Office.
Dd. 289589, C30, 5/88.